Chemical Tests

STUDENT ACTIVITY BOOK

SCIENCE AND TECHNOLOGY FOR CHILDREN

NATIONAL SCIENCE RESOURCES CENTER
Smithsonian Institution • National Academy of Sciences
Arts and Industries Building, Room 1201
Washington, DC 20560

The National Science Resources Center is operated by the Smithsonian Institution and the National Academy of Sciences to improve the teaching of science in the nation's schools. The NSRC collects and disseminates information about exemplary teaching resources, develops and disseminates curriculum materials, and sponsors outreach activities, specifically in the areas of leadership development and technical assistance, to help school districts develop and sustain hands-on science programs. The NSRC is located in the Arts and Industries Building of the Smithsonian Institution and in the Capital Gallery Building in Washington, D.C.

Copyright © 1994, National Academy of Sciences. All rights reserved.
99 98 97 96 95   10 9 8 7 6 5 4 3 2

ISBN 0-89278-705-8

Published by Carolina Biological Supply Company, 2700 York Road, Burlington, NC 27215.
Call toll free 1-800-227-1150.

See specific instructions in lessons and appendices for photocopying.

♻ Printed on recycled paper.   CB787209511

Contents

Thinking about Chemicals

Think and Wonder

You have probably heard the word "chemical" before. What do you know about chemicals? What would you like to learn? Let's talk about these things. Then, you will observe a "mystery bag." What do you think is inside?

Materials

For you
1 science notebook
1 pair of goggles
1 plastic bag

For you and your partner
1 *Chemical Tests* Student Activity Book

For you and your team
1 mystery bag
2 paper towels
1 tray

Find Out for Yourself

1. Put today's date at the top of the first page in your notebook. (Be sure to do this every day you write in your notebook.) In your notebook, record your thoughts about some of these questions.

   ■ What do you know about chemicals?

   ■ Where have you seen chemicals, or what have you heard about them?

   ■ What are some uses of chemicals?

   ■ How have you used chemicals?

2. Share your thoughts with your group. Be prepared to report your group's ideas to the class. Your teacher will keep a list of everyone's ideas.

3. It's time for you and your group to observe a mystery bag with a common chemical inside. Have a group member pick up the materials you will need.

   - 1 tray
   - 2 paper towels
   - 4 pairs of goggles
   - 4 plastic bags for the goggles
   - 1 mystery bag

4. Put the mystery bag on the tray and place the tray in the middle of your group. Keep the paper towels nearby.

5. Take a piece of tape. Write your name on it, and put the tape on your empty plastic bag. Then put on your goggles. If they feel too tight, stretch the band.

6. **Without opening the plastic bag,** observe what is inside. Discover as much as you can about the mystery chemical with your group.

7. Share your group's observations with the rest of the class.

8. How could you find out more about what is inside the bag? Share your ideas with your class.

Figure 1-1

Using your sense of smell

9. Discuss with the class what you have discovered about the mystery chemical so far.

   - What senses have you been using in your observations?
   - What words would you use to describe the mystery chemical?

10. How do you think you could find out even more about the mystery chemical? Discuss your ideas with your class. You and your group will try some of the class's ideas.

11. Clean up. Return the tray, plastic cup, and your goggles (in the plastic bag with your name on it) to the materials center. Throw away the used paper towels and bag of mystery goo.

12. What would you like to learn about chemicals? Share your thoughts with the class. Your teacher will write down your questions.

13. A label has peeled off a container of white powder in your kitchen. In your notebook, describe some ways you can find out what the powder is.

Ideas to Explore

1. Do you have an idea about what the mystery chemical could be? Record your ideas in your notebook and include why you think so. You will find out what the chemical is later in the unit!

2. Start a science journal in one section of your notebook. Write about your science activities in it. Did you make a new discovery today? What surprised you? Keep adding thoughts when you do new activities.

Investigating Unknown Solids: Getting Ready

Think and Wonder

What would it be like to be a detective or scientist—and solve a mystery? Now you can find out. Today, your teacher will give you a mystery to solve. And, like all good science detectives, you will put together some tools you need to start your investigation.

Materials

For you

  1 science notebook

For you and your partner

  1 science pail

  5 jars

  1 plastic bag containing five measuring spoons

  5 small colored dots (red, orange, blue, green, yellow)

10 large colored dots (2 red, 2 orange, 2 blue, 2 green, 2 yellow)

  2 pairs of goggles in their bags

Find Out for Yourself

1. Look and listen as your teacher explains your mystery: What are the identities of the five chemical unknowns?

2. Can you think of some ways to gather information about the unknowns to solve the mystery? Remember how you found out some properties of the unknown chemical in Lesson 1? Be ready to share your ideas with the class.

3. To gather evidence about the unknowns, you will need to put together a science pail of materials. Pay attention as your teacher describes these materials. Discuss with the class why you think safety is important when you are working with different materials and tools in science.

4. Decide with your partner which one of you will pick up your materials. Write your names on the tape on your science pails. Remember, you will store your goggles in the plastic bags.

5. Use the large dots to color-code your five jars and lids.

6. Now you are going to fill your set of five color-coded jars. While you are waiting to fill the jars, put a small dot of a different color on the handle of each measuring spoon. Review the following steps with your teacher.

    When it is your turn to go to the jar-filling station, do the following:

    ■ Put on your goggles.

    ■ With your partner, bring your science pail (with the five sample jars in it) to the jar-filling station.

    ■ Take out the jar whose colored dot matches the colored dot on the container at your workspace.

    ■ Unscrew the lid of your jar. Take the lid off the container at your workspace. Using the spoon for that container, carefully fill your jar.

    ■ When you have finished filling your jar, screw the lid on tightly. Put the jar back in your science pail. Put the lid back on the container at your workspace.

    ■ Wait until the other teams at the station have finished filling their jars. With your partner, move to the next workspace to your right. The other teams will move, too.

Figure 2-1

Filling the unknown sample jars

■ Repeat the same steps. Be sure the colored dot on the jar you are filling matches the dot on the container at your new workspace.

■ When you have filled all five jars, bring your pail back to your regular seat.

■ Now, finish color-coding your measuring spoons.

Figure 2-2

Fully assembled pail

7. After you have filled your jars and color-coded your measuring spoons, look at the picture above. That's what you should have at the end of this activity.

8. Time to clean up. Your teacher will review what to do.

9. Now try a "mystery object" activity to practice observing and describing objects. You and your partner will get a mystery object bag.

   ■ Closely observe the object, but do not take it out of the bag.

   ■ Use properties you have observed to write a brief "What Am I?" description of it. Do not include pictures or tell how the object is used.

   Your teacher has put an example on the chalkboard.

10. After you have written the description, swap descriptions—not mystery object bags—with another team. Read the description. You have one best guess to name the other team's object.

11. Discuss the activity with your class. Your teacher will lead the discussion by asking these questions:

   ■ How many teams guessed the mystery object?

   ■ Why do you think you were able to guess what the object was?

   ■ How could you change the descriptions to help more teams guess the objects?

Ideas to Explore

1. Write "What Am I?" riddles describing your favorite animal or food. Exchange them with a friend. Can he or she guess what you are describing?

2. Create a **cinquain poem** about an object of your choice. To write this kind of poem, use properties you can observe. Below is an example of a cinquain poem. (If you are not sure how to start, ask your teacher for help.)

Figure 2-3

Cinquain poem

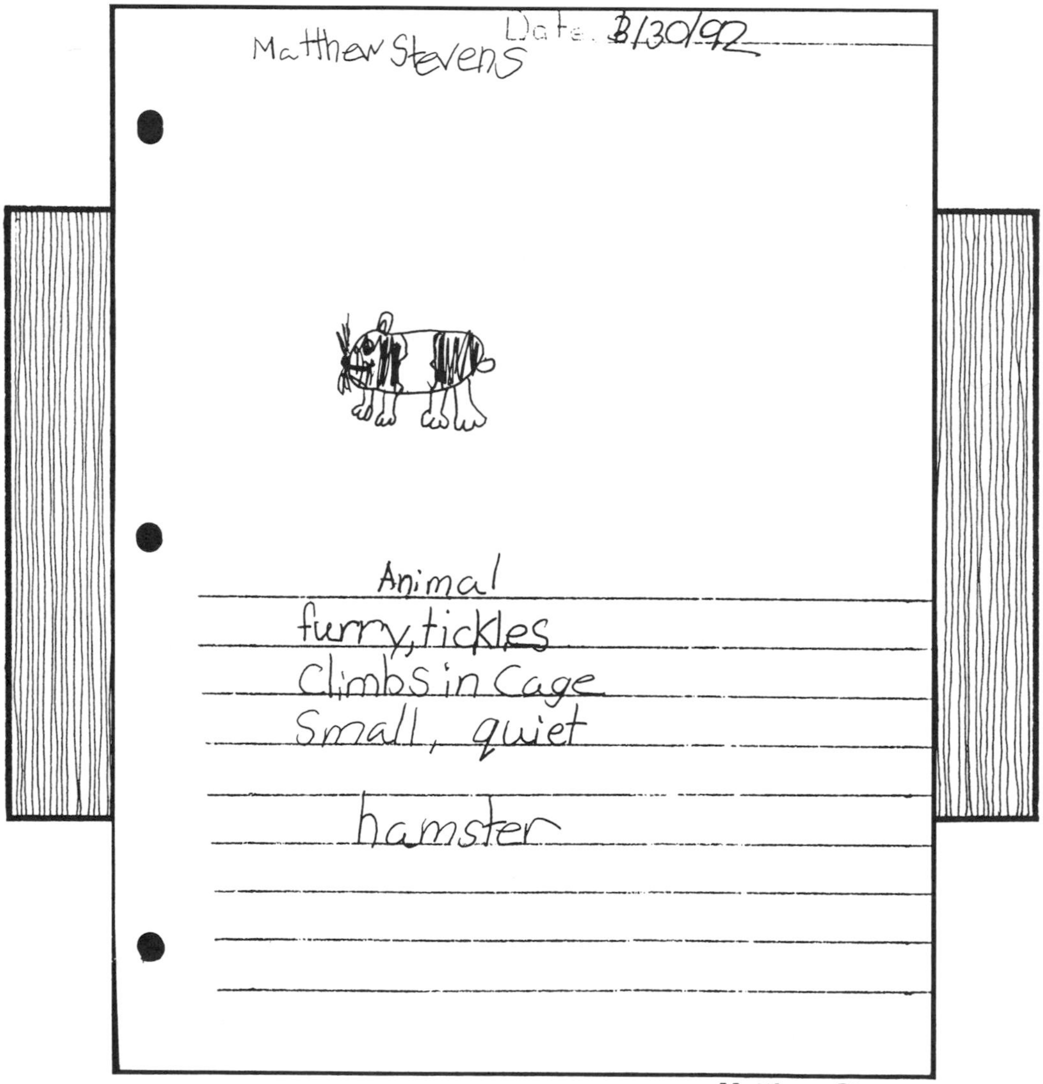

Matthew Stevens   Date 8/30/92

Animal
furry, tickles
Climbs in Cage
Small, quiet

hamster

Matthew Stevens

3. Use black paper to make a silhouette of your partner. Write a cinquain poem, or other descriptive poem, about your partner.

4. Write a description of a mystery object in the classroom. Ask other students to put their guesses into a box or envelope.

Exploring the Five Unknown Solids

Think and Wonder

When detectives investigate mysteries, what they really are doing is exploring unknowns. In this lesson, you will have a chance to observe the five unknown chemicals with your senses (except taste, of course). What can you discover about the unknowns just by using your senses? What else can you use to help you take a closer look at the unknowns?

Materials

For you

1   science notebook

For you and your partner

1   science pail
1   tray
1   test mat
1   sheet of wax paper
5   toothpicks
1   paper towel
2   hand lenses
2   pieces of black construction paper

Find Out for Yourself

1. What is the mystery you are trying to solve? In the last lesson, you shared some ideas on how to find out about the unknowns. Discuss the ideas with your class.

2. Pick up your materials from the materials center.

3. With your teacher, look at the instructions on pg. 15. They show how to set up your tray and take samples of the unknowns.

4. Observe each unknown. As you get samples, take turns with your partner and discuss your observations.

5. Your teacher will give you a hand lens and black construction paper. How can these materials help you observe the unknowns more closely?

6. Using your materials, continue to observe the unknowns. Talk with your partner about the best way to organize your observations. In your notebook, record what you discover about each one.

7. Have you finished? Look at the picture below. It shows you how to fold up the wax paper. Fold it tightly so that it is easy to throw away. Throw away the used toothpicks, too.

Figure 3-1

Cleaning up

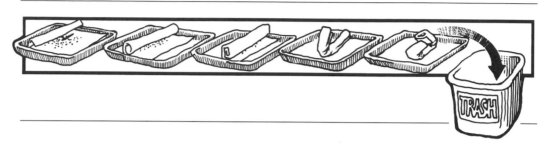

8. Finish cleaning up.

   ■ Check your science pail. Remember, the "Check Your Science Pail" poster will show you what should be in your pail.

   ■ Return the other materials to the materials center, and put your science pail in the pail storage area. Remember to clean up your space.

Figure 3-2

Putting materials back in the science pail

9. Share with the rest of the class the way you organized your observations of the unknowns. Think about these questions:

   ■ Why do you think it might be important to record your observations?

   ■ How could organizing your information help you solve the mystery of the five unknowns' identities?

10. Help your teacher make a "Class Properties Table" of the class's observations of the unknowns. Your teacher will explain what to do.

11. Use the "Class Properties Table" to name a property of each unknown. Share your thoughts with the class.

Ideas to Explore

1. Use a microscope to take an even closer look at each unknown. Make a microscope slide for each unknown. Draw what each sample looks like with your eyes alone and then with the help of the microscope. Compare your drawings. How are they different? How are they alike?

2. Create a **concrete poem** about an object of your choice. In this kind of poem, you use properties both to describe the object **and** to create a picture of it. For an example, look at the picture below. You can also color the picture and add a background.

Figure 3-3

Concrete poems

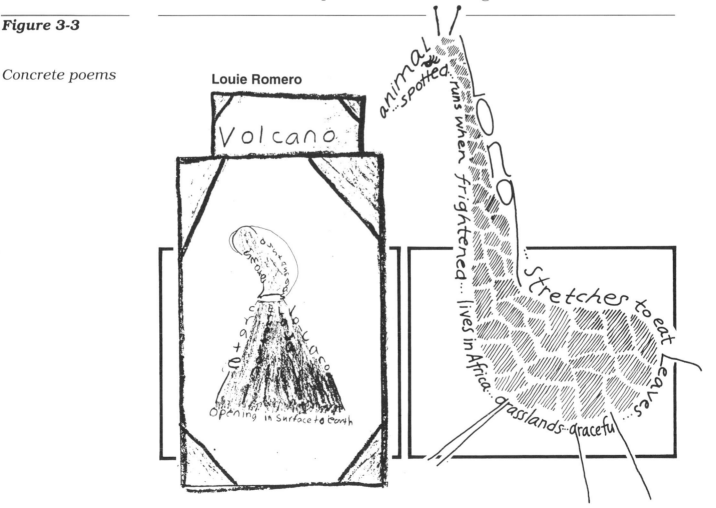

3. Play the "Describer-Guesser" game. Choose an object to describe and write the description, one sentence at a time, for your partner. Your partner will try to guess the object after each descriptive sentence. Try to get your partner to guess the object in as few sentences as possible.

4. Get a rock, a leaf, or another interesting object. List as many properties of it as you can. Ask someone in your family to try to guess what the object is.

Student Instructions for Taking a Sample of the Unknowns

1. Put on your goggles. Set up your tray with the test mat (use the side with the color names on it) and put the wax paper on top of the test mat.

2. Take out the red unknown sample jar and red measuring spoon. Open the jar.

3. Take one measure of the red unknown. Use your toothpick to level the amount on the spoon.

4. Put the sample on the red circle on your test mat.

5. Wipe off the spoon with a paper towel. Put the closed jar and spoon back in your science pail.

6. Observe the red unknown. Repeat Steps 2 through 6 with the four other unknowns.

Testing Unknown Solids with Water

Think and Wonder

In Lesson 3, you observed the properties of the five unknowns. There are some other things you can do to learn more about each unknown. Let's find out what they are.

Materials

For you

   1   science notebook

   1   **Record Sheet 4-A: Test Results Table**

For you and your partner

   1   science pail

   1   dropper bottle of water

   1   tray

   1   test mat

   2   paper towels

   1   sheet of wax paper

   5   toothpicks

Find Out for Yourself

1. Look at the "How We Are Finding Out about the Unknowns" list. What do you think the next step in your investigation should be? Discuss this with your class.

2. What do you think you might observe if you added a few drops of water to each unknown? Record your predictions in your notebook.

3. When you get **Record Sheet 4-A: Test Results Table,** pay attention as your teacher reviews how to set up and label your test table.

4. Listen as your teacher reviews the instructions on pg. 20. They explain how to do the water test.

5. Pick up your materials for the water test. Look again at the instructions on pg. 20 if you need help.

Figure 4-1

Folding and labeling your test table

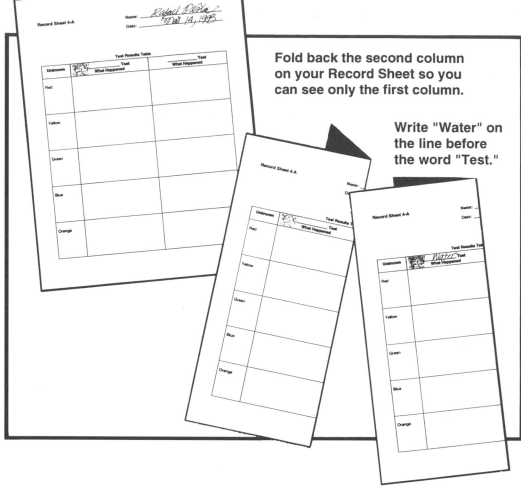

Fold back the second column on your Record Sheet so you can see only the first column.

Write "Water" on the line before the word "Test."

6. Clean up your materials and workspace. Look back at Lesson 3 for cleanup reminders.

7. Write in your notebook what you found out by adding water to the unknowns.

8. Share your results with your class. Use Record Sheet 4-A to help you.

9. Compare the predictions in your notebook with your test results. Were you surprised by your results? Share your thoughts with the class.

10. What differences did you notice among the reactions? What do you know now about each unknown that you didn't know before? Talk about this with your class.

11. Put Record Sheet 4-A in your notebook to use for the next lesson.

12. Your teacher will show you the "Class Prediction Table," on which you can record what you think the unknowns may be and why you think so. Record your ideas in your notebook too.

**Ideas to
Explore**

1. With your teacher and classmates, sit in a circle on the floor. Along with them, place one of your shoes in the middle of the circle. Watch your teacher closely as he or she begins to sort the pile by using one property of the shoes. Which property is your teacher using to sort the shoes? After all shoes have been put back in one pile, watch while a student sorts them by a new property.

2. What happens when you put a drop of food coloring in water? Try it. Get a clear cup of water and then put a drop of food coloring in the cup. Watch closely and describe what you see.

Student Instructions for Doing the Water Test

1. Set up your tray, test mat, and wax paper. With the red measuring spoon, take one sample of the red unknown. Use your toothpick to level the amount on the spoon. Put the unknown in the red test circle.

2. Wipe off the measuring spoon with a paper towel before you put it back in the spoon bag. Put the unknown jar back in your science pail.

3. Put six drops of water on the red unknown sample. What happens? Use your hand lens for a closer look.

4. Use a toothpick to mix the water with the unknown. (Be careful not to rip the wax paper.) Now what happens? Record your observations on the water test table.

5. Repeat Steps 1 through 4 for the four other unknowns.

Exploring Water Mixtures

Think and Wonder

In Lesson 4, you discovered what happens when you add a few drops of water to each unknown. What do you think will happen when you mix each unknown with a lot of water? And, after you mix them, can you take them apart? If so, how?

Materials

For you

    1 science notebook

    1 **Record Sheet 4-A: Test Results Table** (from Lesson 4)

For you and your partner

    1 science pail

    1 tray

    1 test mat

    5 small plastic cups (red, orange, blue, green, yellow)

  10 small colored dots (2 red, 2 orange, 2 blue, 2 green, 2 yellow)

    1 plastic dropper

    1 large plastic cup

    2 paper towels

    1 sheet of wax paper

    5 toothpicks

Find Out for Yourself

1. What do you predict will happen if you mix each unknown with more than six drops of water? Record your thoughts in your notebook.

2. Take out **Record Sheet 4-A: Test Results Table** from your notebook. Look at the second column of the table, where you will record today's observations. Write "Water Mixtures" on the blank line before the word "Test" at the top of the column.

3. Pick up your materials from the materials center.

4. Listen while your teacher goes over the instructions on pg. 25 for doing the water mixtures test. Then you can begin testing.

5. When everyone has finished the testing, share your observations with your class.

6. Compare the predictions in your notebooks with your test results. Share your findings with the class.

7. Now that you have mixed the unknowns with water, do you think you can separate them again? Discuss your ideas with your class. Put Record Sheet 4-A back in your notebook.

8. You and your partner will now try to separate the unknowns from the water. The instructions on pg. 27 will show you how. Listen as your teacher goes over the steps.

9. After your teacher gives you your materials, do the filtration activity. Your teacher will show you where to put your tray of dishes and the five wet filters.

10. Discuss with your class what you observed during filtration.

11. Your teacher will tell you where to take the small plastic cups. Then clean up your other materials and workspace.

12. Record in your notebook some thoughts about the following questions:

    - What did you learn by creating mixtures of the unknowns and water?

    - In what ways were the five mixtures similar?

    - In what ways were the mixtures different?

    - What were some properties of each mixture?

Ideas to Explore

1. Get a quart jar (mayonnaise jars work well) and create "dirty" water with soil, sticks, and leaves. Then design a method to clean the dirty water. Describe it in your notebook. You can use pictures, words, or both.

2. Have you ever wondered where the water that comes out of the faucet at school comes from? Or how dirty water is cleaned? Read *The Magic School Bus at the Waterworks,* by Joanna Cole, and find out!

3. Are you curious about the other units of measurement on the small plastic cup? Find out why there are different kinds of units. How is each kind used today?

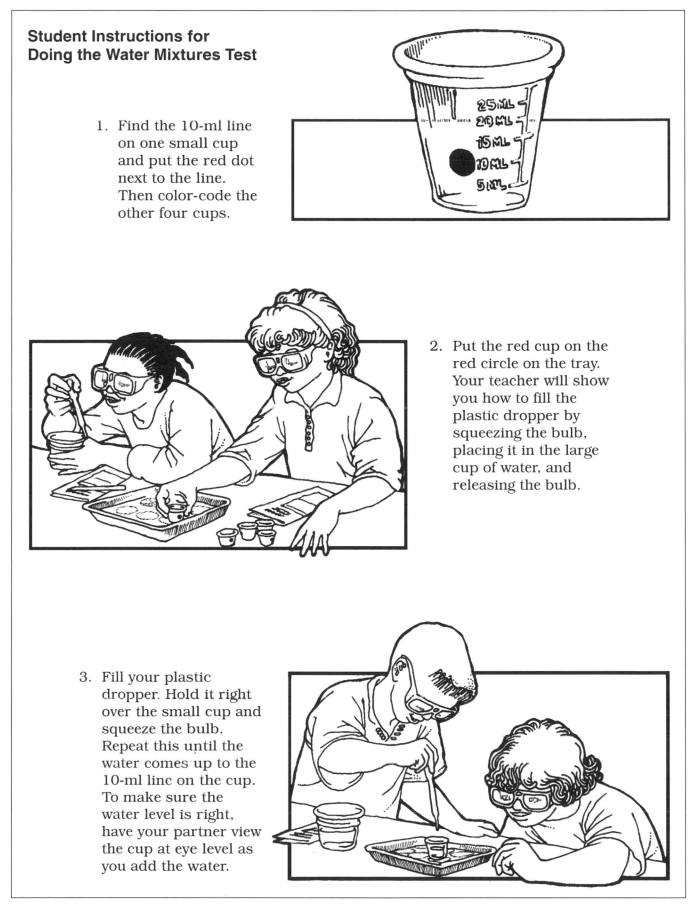

Student Instructions for Doing the Water Mixtures Test

1. Find the 10-ml line on one small cup and put the red dot next to the line. Then color-code the other four cups.

2. Put the red cup on the red circle on the tray. Your teacher will show you how to fill the plastic dropper by squeezing the bulb, placing it in the large cup of water, and releasing the bulb.

3. Fill your plastic dropper. Hold it right over the small cup and squeeze the bulb. Repeat this until the water comes up to the 10-ml linc on the cup. To make sure the water level is right, have your partner view the cup at eye level as you add the water.

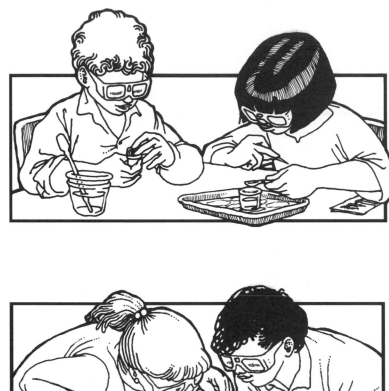

4. Using your red measuring spoon, add one measure of the red unknown to the cup. Stir the mixture with a toothpick for 30 seconds.

5. Add two more measures of the red unknown **one at a time.** After each measure, stir again for 30 seconds. (You now have added a total of three measures.)

6. Observe the mixture you have just made.

7. Let the cup sit on the tray. Repeat Steps 2 through 6 for the four other unknowns. Switch jobs with your partner so that you each get a chance to use the dropper.

8. Record on **Record Sheet 4-A** short answers to the following questions:

   ■ Where is the unknown in the water? Can you still see it?

   ■ What does the water mixture look like? Compare it with the plain water left in your large cup. Has it changed? If so, how?

   ■ How does stirring affect the way the mixture looks?

Student Instructions for Filtering the Water Mixtures

1. With your partner, use a pencil to write your names near the top of each filter. Under your names, label one filter red, one orange, one green, one blue, and one yellow. Then color-code the dishes by placing one dot on the side of each dish.

2. Carefully move the small cups from the tray to your desk. Make sure to keep the cups away from the edge of the desk and away from your arm.

3. Put the red dish on the tray. Pick up the red filter and use both hands to hold it open directly over the dish.

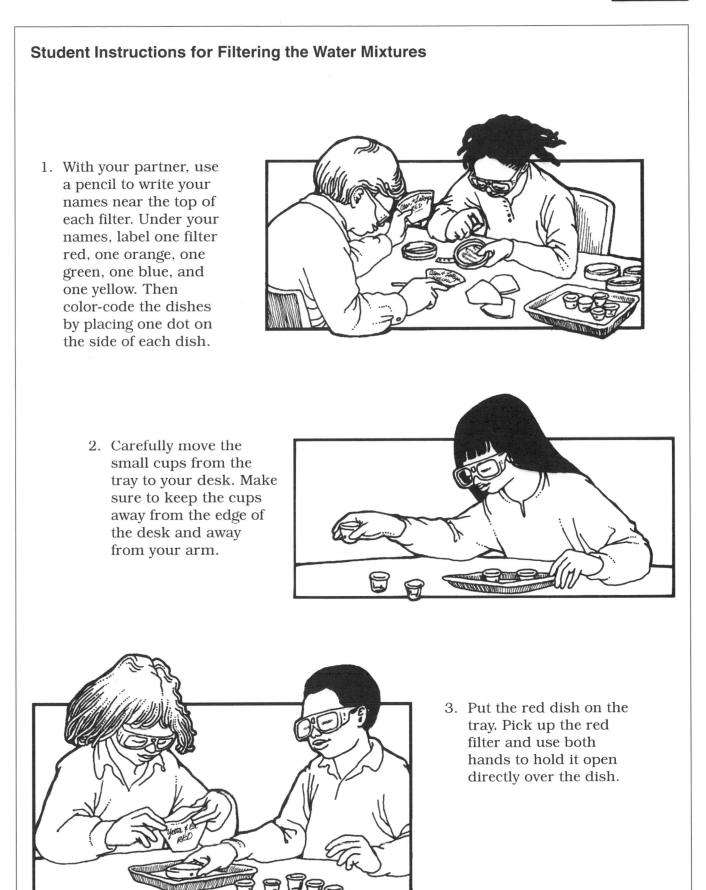

4. Have your partner stir the water mixture in the red cup once with the toothpick and then slowly pour it into the filter. Observe what happens.

5. Wait until no more liquid is dripping out of the filter. Leave the dish undisturbed on the test mat. Put the filter near the tray on your desk.

6. Repeat Steps 3 through 5 for each mixture.

Discovering Crystals

Think and Wonder

What happened to the water mixtures you filtered in Lesson 5? Are the results what you expected? Let's find out.

Materials

For you

   1 science notebook

   1 **Record Sheet 6-A: Growing Sugar Crystals**

   1 **Record Sheet 6-B: Growing a Giant Crystal**

For you and your partner

   1 tray with five dishes (from Lesson 4)

   5 filters (from Lesson 4)

   1 science pail

Find Out for Yourself

1. Take out your notebook and turn to the filtration results table you made.

2. What did you notice about the filters and dishes while they were left to sit in the classroom? Share your observations with your class.

3. Pick up your tray of dishes, five filters, and science pail. Your teacher will pair you up with another team for the next step.

4. Review the steps below with your teacher.

   - Observe the dish and open the filter paper for each unknown.

   - Record your results for each unknown on the filtration results table in your notebook.

   - Compare what is left in the filters and dishes with the untested samples in the unknown jars. Have the unknowns changed? In what ways?

Figure 6-1

*Observing your
filtration results*

- Discuss your results with another team. Talk about the answers to the following questions:

  - What happened to the mixtures after filtration?

  - Are your results similar to those of the other team?

  - If your results are different, why do you think that happened?

  - Which results surprised you? Why?

5. Now, follow the above steps.

6. Share your results in a class discussion.

7. Discuss your water mixture test and filtration results with your class. What do you know about the unknowns now that you did not know before?

8. Put the dishes in the soapy bucket when your teacher tells you. Clean up the rest of your materials and throw away the filters.

9. Read the story about crystals.

10. Write in your notebook three new things you learned about crystals from reading the story. Also write down any questions you still have.

11. Your teacher will give you two crystal activities to do at home. Choose one of them. After you have completed the activity, you will share your results with the class.

Ideas to Explore

1. Using the crystals your class grew, compare the red unknown crystals with the yellow unknown crystals. How are they alike? How are they different? Bring samples of crystals from home to share with your class.

2. Read *Two Bad Ants,* by Chris Van Allsburg. It's the story of how crystals changed the lives of two ants.

3. With your class, find out how silicon chips help run computers and VCRs. One way would be to write a letter to a computer company. What are some others?

Reading Selection

A World of Crystals

Did you know that crystals are all around you? Outside, they make up rocks, minerals, snow, and the sand that crunches between your toes at the beach. Inside, your home is filled with crystals. In your freezer are ice crystals. In your Mom's jewelry box, there may be gems or metals that are crystals. You sprinkle crystals of sugar on your cereal and mix crystals with water to make lemonade.

A sparkling diamond. A snowflake. A speck of sugar or a grain of salt. These are all crystals. They seem like very different things. But in some ways, they are the same. How? All crystals are solid. They are made up of pieces. And those pieces are laid out in a pattern. That pattern is repeated over and over again.

All crystals of the same material have a similar shape. Take quartz, for example. You can look at a piece of quartz the size of a

Quartz crystal

pea or one the size of a pumpkin. The quartz can be from Africa—or from right here in the United States. No matter what, the shape of the quartz crystal is always similar.

Cubes on Your French Fries?

Some crystals are so small that you must look through a microscope to see them. Then you'd know their shapes. Did you know you were eating tiny cubes on your french fries? Or that if you could catch a snowflake and see it under a microscope, it would have six points?

Thanks to a famous scientist named Roger Bacon, crystals help us take a closer look at the world. A long time ago, Bacon discovered that by shaping and polishing quartz, he could make a lens. This lens made small objects look bigger. And that was the start of modern eyeglasses, microscopes, telescopes, and even contact lenses.

Today, it's hard to imagine living without these things. But back then, people thought that lenses were some kind of black magic and that Roger Bacon was a wizard. They actually put him in jail for his important discovery.

Snowflakes

Many Uses

Crystals are valuable in other ways. For ages, kings and queens have worn crowns made of precious metals such as gold and silver. Pirates lined their chests with jewels such as red rubies, green emeralds, and diamonds.

In fact, diamonds are not only beautiful to look at, but they are also very useful. Because they are so hard, they can be used to saw, drill, grind, and polish. So, we use diamond saws to cut microchips for computers. We use diamond dust to polish glass and other hard things.

Today, we have many more uses for crystals, especially a part of quartz called "silicon." Silicon helps run and control computers, calculators, and microwave ovens. Also, quartz helps make some watches work.

Crystals are a mystery, in a way. They are not alive, and yet they grow. Want to see how? Ask your teacher about how you can grow some crystals at home.

Silicon helps run computers, microwaves, and watches.

Testing Unknown Solids with Vinegar

Think and Wonder

Now that you have investigated and recorded some of the properties of the five mystery chemicals, are you ready to test for new properties? Before, you added water. Now, you will add vinegar. What new properties will you discover?

Materials

For you

- 1 science notebook
- 1 **Record Sheet 7-A: Test Results Table**

For you and your partner

- 1 science pail
- 1 dropper bottle of vinegar
- 1 tray
- 1 test mat
- 1 sheet of wax paper
- 5 toothpicks
- 1 paper towel

Find Out for Yourself

1. What do you need to add to the class list "How We Are Finding Out about the Unknowns"? Tell your teacher.

2. What do you know about vinegar? What do you think might happen if you added a few drops of vinegar to each unknown? Record your predictions in your notebook.

3. When you get **Record Sheet 7-A: Test Results Table,** fold back the second test table column as you did in Lesson 5. At the top of the first column, write "Vinegar" in front of the word "Test."

4. Review with your teacher the instructions for the vinegar test on pg. 37.

5. Pick up your materials and follow the instructions for the vinegar test.

6. Clean up your materials and workspace. Remember to look at the "Check Your Science Pail" poster.

7. Share your vinegar test results with your class and teacher. Think about these questions:

   ■ How did the vinegar react with each unknown? In what ways were the reactions similar? Different?

   ■ How did the untested sample in the compare circle help you describe your observations?

   ■ How were the vinegar test results similar to those of the water test? How were they different?

   ■ Did any unknowns react differently from the others?

   ■ How did your predictions compare with your results?

8. Record in your notebook what you have learned by adding vinegar to the unknown.

Ideas to Explore

1. Have you ever made a picture using colored powders and crystals? If your teacher says it's okay, make powdered-object mosaics. You need glue, salt, sand, nondairy creamer, food coloring, and dark construction paper.

   ■ Color the salt, sand, and creamer with different food colorings.

   ■ Use lines of white glue to design a pattern or scene on the paper.

   ■ Now sprinkle the colored salt, creamer, and sand on the glue lines wherever you choose.

   ■ How did the properties of the materials create a mosaic? Discuss this question with your class.

2. Try some kitchen science activities at home. You can get good ideas from books such as *Science Experiments You Can Eat,* by Vicki Cobb.

3. Ask your teacher if you can create a winter scene. Cut out shapes from colored construction paper. You can add details with crayons. Glue the shapes onto a dark piece of paper. Next, your teacher will give you some solution. Use a paintbrush to cover the picture with the solution. Let it dry thoroughly. How did the solution change the picture? Describe this to your teacher.

Student Instructions for Doing the Vinegar Test

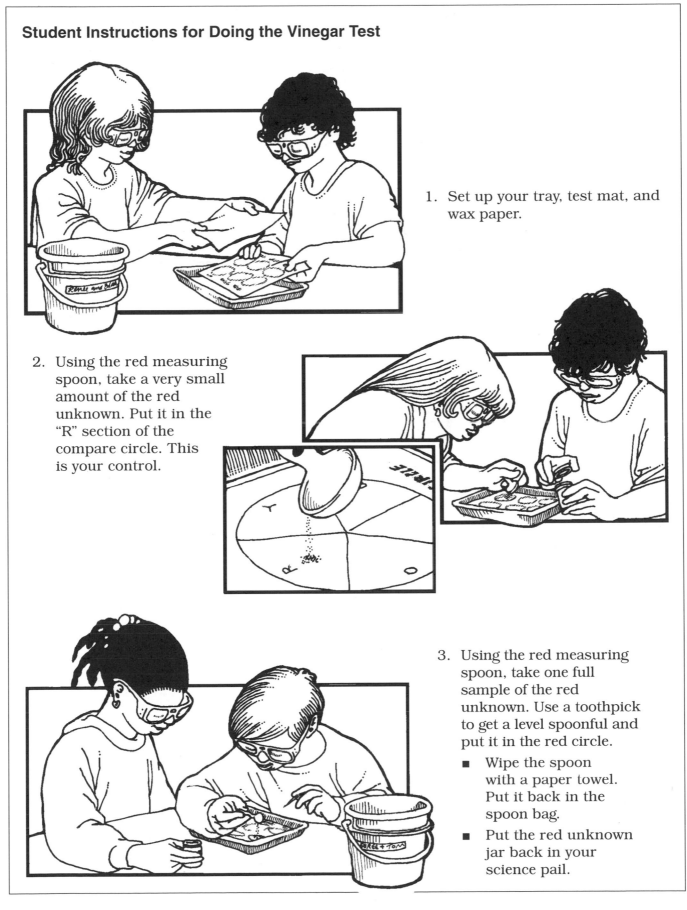

1. Set up your tray, test mat, and wax paper.

2. Using the red measuring spoon, take a very small amount of the red unknown. Put it in the "R" section of the compare circle. This is your control.

3. Using the red measuring spoon, take one full sample of the red unknown. Use a toothpick to get a level spoonful and put it in the red circle.

   ■ Wipe the spoon with a paper towel. Put it back in the spoon bag.

   ■ Put the red unknown jar back in your science pail.

4. Put six drops of vinegar on the sample of red unknown in the red circle.

■ Wait a few seconds. What happens?

■ Look at the sample in the compare circle. Compare it with the unknown in the red circle.

■ Did the unknown change after vinegar was added? If so, how? Record your observations on your "Test Results Table."

5. Use a toothpick to mix the vinegar and the red unknown. What happens now? Record any new observations.

6. Repeat Steps 2 through 5 for the four other unknowns.

Testing Unknown Solids with Iodine

Think and Wonder

Now you have observed what happens when you mix one chemical (vinegar) with the five unknowns. There's another chemical you may recognize: iodine. What do you think might happen if you test the unknowns with iodine?

Materials

For you
- 1 science notebook
- 1 **Record Sheet 7-A: Test Results Table** (from Lesson 7)

For you and your partner
- 1 science pail
- 1 dropper bottle of iodine
- 1 tray
- 1 test mat
- 5 toothpicks
- 1 sheet of wax paper
- 1 paper towel

Find Out for Yourself

1. With your teacher, add the vinegar test to the class list "How We Are Finding Out about the Unknowns."

2. What do you know about iodine? What do you think might happen if you added iodine to each of the unknowns? Record your predictions in your notebook and then share them with the class.

3. Take out **Record Sheet 7-A: Test Results Table.** Unfold it and write "Iodine" at the top of the second column in front of the word "Test." Remember to record your observations for the iodine test on this table.

4. Pick up your materials and do the iodine test. You can look back at Lesson 7, pg. 37, if you need to reread the instructions.

5. Clean up your materials and workspace.

Figure 8-1

*Clean up all
spills right away*

6. Share your iodine test results with your class. Did any unknown react differently from the others?

7. How did your predictions compare with your results? Did any results surprise you? Share your thoughts with the class.

8. In your notebook, record answers to these questions:

   ■ What have you learned about the unknowns by doing the iodine test?

   ■ Look at all your test results so far. Have any of the unknowns reacted similarly when you used different tests?

**Ideas to
Explore**

1. With your teacher, place a drop of iodine on a microscope slide and spread it out with a toothpick. Allow it to dry. Observe the slide under a microscope. Breathe on the slide. What happens to the crystals?

2. What do you think makes a good mystery? What do you like about the mystery books you have read? Write your own mystery story. Try to put the things you like in your story.

Testing Unknown Solids with Red Cabbage Juice

Think and Wonder

Are you keeping count? So far, you have done four chemical tests on the unknowns. With each test, you are making new observations that will bring you closer to solving your mystery. Now, let's use a liquid to do one more chemical test. This time, you will use the juice of a plant.

Materials

For you

1   science notebook
1   **Record Sheet 9-A: Test Results Table**

For you and your partner

1   science pail
1   dropper bottle of red cabbage juice
1   tray
1   test mat
5   toothpicks
1   sheet of wax paper

Find Out for Yourself

1. What do you know about red cabbage or red cabbage juice? What do you think might happen if you added red cabbage juice to the unknowns? Record your ideas in your notebook. Then discuss them with the class.

2. Your teacher will give you **Record Sheet 9-A: Test Results Table.** Fold back the second column and write "Red Cabbage Juice" before the word "Test" at the top of the first column.

3. To do this test, follow the same steps as you did with the vinegar and iodine. The instructions are in Lesson 7, pg. 37. After you add the juice, observe each unknown for about 15 seconds. Then record your observations.

4. Pick up your materials. Do the test.

Figure 9-1

Observe the unknown for 15 seconds before you record your results

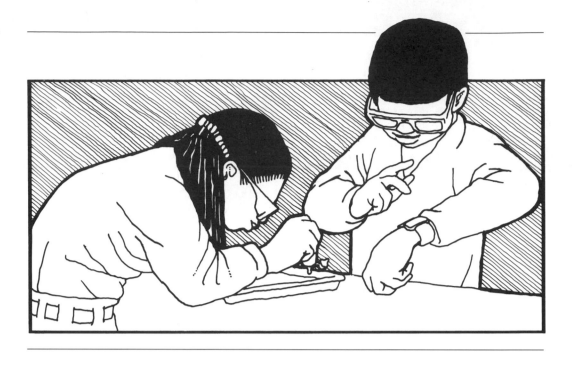

5. Clean up your materials and workspace. Remember to look at the "Check Your Science Pail" poster.

6. Share your test results with your class. Did any unknowns react differently from the others? How did the compare circle help you describe your observations? How did your predictions compare with your results?

7. With your teacher, add the iodine and red cabbage juice tests to the "How We Are Finding Out about the Unknowns" list. Help the class review all the tests you have done so far.

8. In your notebook, record your thoughts about these questions:

   ■ What do you know now about each unknown that you did not know before?

   ■ In what ways are the unknowns similar? In what ways are they different?

   ■ What do you know now about chemicals that you did not know before? What new questions do you have?

9. Next, look at the "What We Know about Chemicals" list. Now that you have done four chemical tests on the unknowns, what can you add to this list? Share your ideas with your class.

10. What questions would you like to add to the "What We Would Like to Know about Chemicals" list? Tell your teacher.

Figure 9-2

Sharing results

Ideas to Explore

1. Did you know you can use plant juices to make paints? Read a book such as *The Legend of the Indian Paintbrush*, by Tomie DePaola. See whether you can find out some of the plants Native Americans use to dye cloth.

2. Have you ever tasted red cabbage? Look in a cookbook to see whether you can find some recipes that use red cabbage.

Testing Unknown Solids with Heat

Think and Wonder

Have you ever heated soup or hot chocolate? What happens to them? What do you think might happen if you heat the unknowns?

Materials

For you

   1 science notebook
   1 **Record Sheet 9-A: Test Results Table** (from Lesson 9)
   1 pair of goggles

For you and your team

   5 aluminum bake cups
   1 foil-lined tray
   5 wooden clothespins
   1 aluminum candle holder
   1 pie pan with water
   1 candle
   5 toothpicks
   5 jars of unknowns
   5 measuring spoons

Find Out for Yourself

1. Have you seen things heated in a kitchen or over a campfire? What do you think might happen if you heat the unknowns? Record your predictions in your notebook.

2. Take out **Record Sheet 9-A: Test Results Table**. Unfold it and write "Heat" at the top of the second column in front of the word "Test."

3. Why do you think safety is important when you heat something? Help your teacher add some safety rules for heating to the "Safety Rules" list. Get your goggles from your science pail.

4. Review the heat test steps on pg. 49 with your teacher. This is what your group will do when it's your turn at the heat station.

5. After your group performs the heat test, take your notebook to the follow-up station to discuss the test results. The group helper will bring your group's tray of aluminum bake cups.

6. With your group, discuss what you observed during the heat test. For each unknown, talk about the following questions:

   ■ What did you observe about the unknown when it was heated?

   ■ What do you observe about the unknown after it has cooled?

   ■ In what ways has the unknown changed as a result of heating?

   The group helper should throw away the aluminum cups and return the tray and clothespins to the materials center.

7. Put your test table in your notebook. Return to your seat and work on the other activities your teacher has assigned. Keep your science notebook at your desk.

8. After all the groups have been to the heat station, share your heat test results with the rest of the class. How did your predictions compare with your results?

9. What did you discover from heating the unknowns that you did not learn from adding liquids to them? Record your ideas in your notebook.

Ideas to Explore

1. Have you ever wondered what causes the colors in fireworks? Go to the library to find out.

2. What causes the colors in fluorescent lights? Do some research to find this out.

Student Instructions for Doing the Heat Test

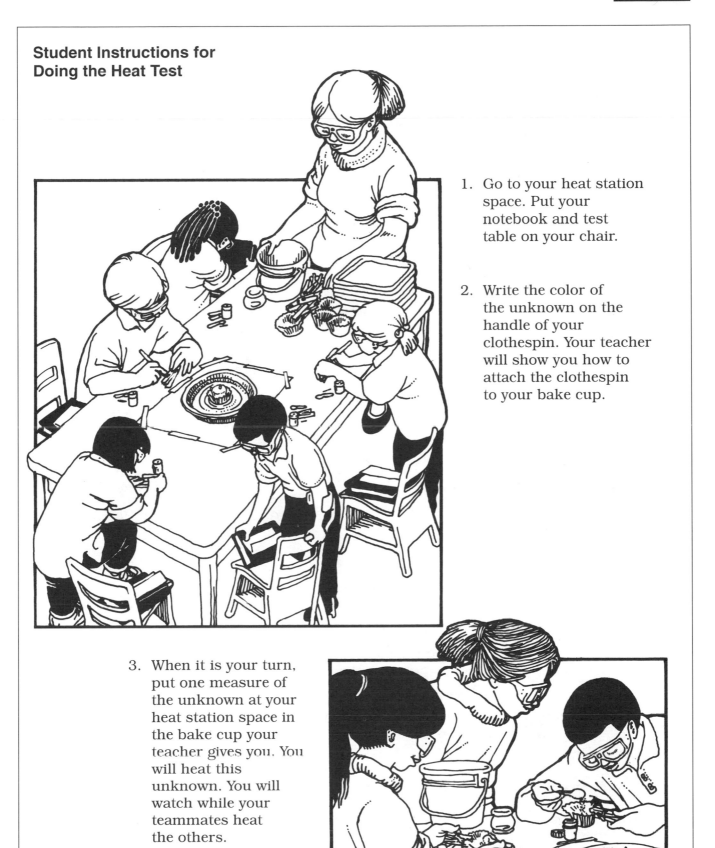

1. Go to your heat station space. Put your notebook and test table on your chair.

2. Write the color of the unknown on the handle of your clothespin. Your teacher will show you how to attach the clothespin to your bake cup.

3. When it is your turn, put one measure of the unknown at your heat station space in the bake cup your teacher gives you. You will heat this unknown. You will watch while your teammates heat the others.

4. When heating an unknown, hold the cup one inch above the flame. Make sure the sample in the bake cup is directly over the flame.

5. When you observe the unknown as it is being heated, do not lean directly over the unknown.

6. Without leaning on the table, record your test results. Repeat these steps for each unknown.

7. When you are finished, carefully place your bake cup on your team's tray.

Reviewing the Evidence

Think and Wonder

You have observed the unknowns and performed six tests on them. Now you will put together all the evidence you have collected. How can you use this information? These clues can help you solve the mystery of the unknowns.

Materials

For you

   1   science notebook (containing **Record Sheets 4-A, 7-A,** and **9-A**)

   1   large piece of white construction paper

   1   scissors

For you and your partner

   Glue

Find Out for Yourself

1. From your notebook, take out **Record Sheets 4-A, 7-A,** and **9-A.** These show the results of the six tests you did on the unknowns.

2. How can you use your test results to help you identify the unknowns? Share your ideas with your class.

3. Look at the "Class Properties Table." Would it help to make a big table like this showing the results from all six tests? Why? Discuss your thoughts with the class.

4. Review the steps for making a "Test Summary Table" with your teacher.

5. Pick up your materials and get to work.

6. Look at the vinegar test results recorded on your "Test Summary Table." Which unknown(s) revealed especially interesting results. Circle them or put a star by them. Share your choice with your teacher and class.

7. Use a pencil or marker to circle or put a star next to the other test results you think were especially important. Which ones helped you tell one unknown from the others?

Figure 11-1

Putting together the "Test Summary Table"

1. **Trim the left margin of Record Sheet 4-A (water and water mixtures). Glue Record Sheet 4-A onto the far left side of the construction paper.**

2. **Take Record Sheet 7-A (vinegar and iodine). Cut off the column of color names.**

3. **Glue Record Sheet 7-A onto the construction paper, next to the water and mixtures table. Be careful to align the columns.**

4. **Take Record Sheet 9-A (cabbage juice and heat). Cut off the column of color names, and trim the right margin.**

5. **Glue Record Sheet 9-A next to the vinegar and iodine table. The right side will hang off the paper a little.**

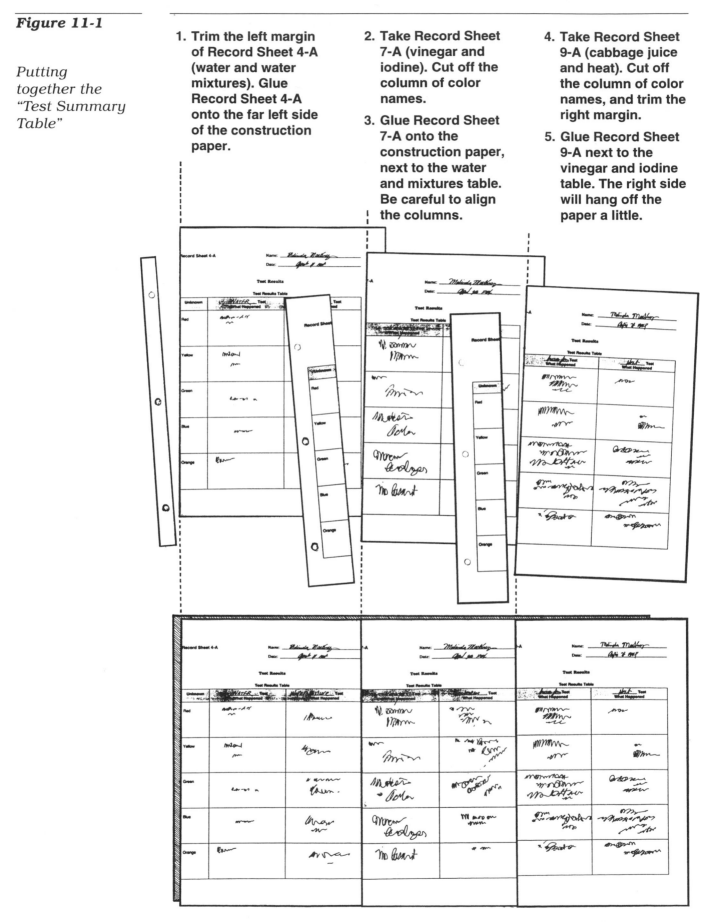

8. Now that you have highlighted the results you think are most important, how can you use them to solve the mystery of the unknowns? Share your ideas with the class.

9. Volunteer to tell the class which test results you highlighted and why.

10. Fold your "Test Summary Table" and put it in your notebook.

Identifying the Unknown Solids

Think and Wonder

You have performed many tests and collected lots of evidence. Like any good detective, you are ready to use that evidence to solve the mystery of the five unknown chemicals!

Materials

For you

   1 science notebook

   1 **Record Sheet 12-A: Chemical Information Sheet**

   1 **Reading Selection: "Chemicals Are All around Us"**

Find Out for Yourself

1. Take out your "Test Summary Table" from Lesson 11. How might you use this information to identify the unknowns? Share your thoughts with the class.

2. Use your "Test Summary Table," the **Chemical Information Sheet** your teacher gives you, and the "Class Properties Table" to help discover the identities of the unknowns.

3. On the Chemical Information Sheet, fill in the color for each unknown under its matching description.

4. You have solved the mystery of the five unknowns! Discuss your findings with your class.

5. In your notebook, write your thoughts about these questions:

   ■ Did any of the unknown's identities surprise you? If so, why? If not, why not?

   ■ How did your predictions of the unknowns' identities compare with your discoveries?

   ■ Which test results were the best clues in helping you identify a specific unknown?

6. With your class, read and discuss "Chemicals Are All around Us."

Figure 12-1

Discovering the identities of the unknowns

7. What would it be like to live in a land that did not have one or more of the chemicals you have just read about? Write a story and then illustrate it.

8. Put your "Test Summary Table" and Chemical Information Sheet in your notebook.

Ideas to Explore

1. Create an "Everyday Chemicals" bulletin board with your class. Draw or bring in pictures of common chemicals. Find out their uses and share them with your class.

2. It's very important to store and handle chemicals safely. Discuss the safe use of household chemicals with your teacher or another adult.

3. Remember the crystals you grew in Lesson 6? How can you prove that they are the red unknown and the yellow unknown?

Identifying the "Mystery Bag Chemical"

Think and Wonder

Do you remember the "mystery goo" from Lesson 1? What do you think it could be? You now have all the information you need to discover its identity. So, now you'll solve that mystery, too!

Materials

For you

- 1 science notebook
- 1 **Record Sheet 13-A: Mystery Goo Test Results Table**

For you and your partner

- 1 science pail
- 5 toothpicks
- 1 tray
- 1 test mat
- 1 sheet of wax paper
- 1 bag of mystery goo
- 1 dry-erase marker

Find Out for Yourself

1. Now that you have solved the mystery of the five unknowns, with your class go over the "How We Are Finding Out about the Unknowns" list. What steps are missing?

2. How can you discover what chemical is in the mystery bag from Lesson 1? Discuss this with your class.

3. Review **Record Sheet 13-A: Mystery Goo Test Results Table** with your teacher.

4. Pick up a bag of mystery goo from the materials center. Observe the goo with your partner.

5. What do you think the chemical is? Base your prediction on your observations and information in your notebook. Record your prediction in your notebook.

Figure 13-1

Labeling the test circles and testing the goo

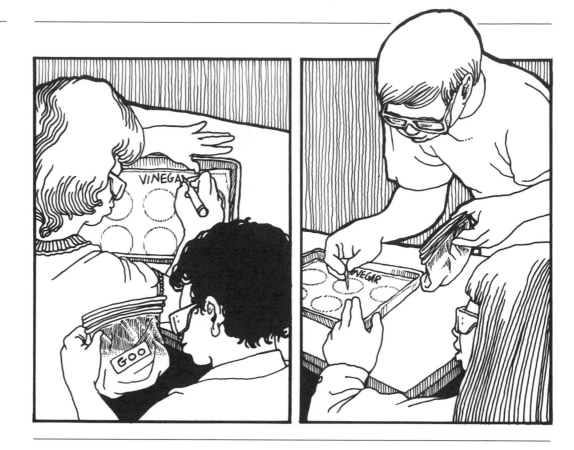

6. Go over the testing instructions on pg. 59.

7. Pick up your materials. Test the goo.

8. Clean up. Throw away the bag of goo.

9. What did you discover? Share your answers to the following questions with the class:

   ■ Which tests did you choose to do? Why?

   ■ Which test result gave you the most information about the properties of the unknown chemical? Why?

   ■ What information in your notebook did you find the most useful?

   ■ How did your prediction about the identity of the goo compare with your results?

10. Share with the class what you think the mystery goo is and why you think so. Your teacher will collect your record sheet.

Ideas to Explore

1. Your teacher may give you a logic problem. Try it. Then create your own logic problem and ask a friend to solve it.

2. Read a mystery book. How do the main characters solve their problem? What steps do they take?

Student Instructions for Testing the Mystery Goo

1. Set up your materials. Use the side of the test mat with the unlabeled circles. Cover it with wax paper.

2. Using one of your measuring spoons, put a sample of the goo inside the compare circle.

3. With your partner, decide which test you want to do first. Use the marker to write the test's name on the wax paper above one of the circles. Also record the test's name in the "What I Did" column of **Record Sheet 13-A.**

4. Put a sample of the goo in the circle you have just marked. Test the goo. Record your observations in the "What Happened" column of the record sheet.

5. Decide on another test. Then repeat Steps 3 and 4.

6. Continue testing and using the information in your notebooks until you think you know which of the five unknowns the goo contains. Fill in the "What I Think It Is" box on the record sheet. Then write in at least two reasons in the "Why I Think So" box.

Testing Mixtures of Two Unknown Solids

Think and Wonder

Are you ready for a new mystery? A jar contains a mixture of two of the five chemicals you have been working with. Which two chemicals are they? How can you find out?

Figure 14-1

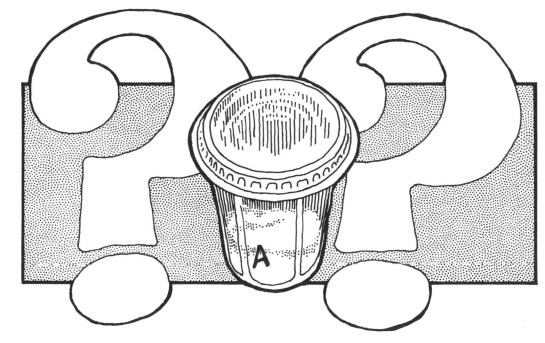

Materials

For you

    1   science notebook

    1   **Record Sheet 14-A: Unknown Mixtures Test Results Table**

For you and your partner

    1   science pail

    1   tray

    1   test mat

    1   sheet of wax paper

5   toothpicks
1   cup of unknown mixture (A, B, or C)
1   dry-erase marker

Find Out for Yourself

1. Look at the cups your teacher holds up. How could you find out which of the five chemicals you worked with are in each mixture? Share your ideas with your class.

2. Go over **Record Sheet 14-A: Unknown Mixtures Test Results Table** with your teacher. What are some ways you might test the unknown mixtures? Discuss them with your class.

3. Pick up your materials.

4. When you and your partner get your unknown mixture, record on your record sheet whether you have A, B, or C.

5. Now start your testing. For help with the testing steps, you can look at pg. 63.

6. Clean up your materials.

7. Share your test results with your class. What chemicals do you think your unknown mixture contains? Why do you think so?

8. Hand in your record sheet to your teacher.

9. What people might answer a question or solve a problem using a method similar to the one you used to identify the unknown mixture? Your teacher will list your ideas.

Ideas to Explore

1. Ask your teacher if the class can invite a detective, scientist, or mystery writer to discuss his or her work. Be sure to explain to this person what you have been doing in this unit.

2. Did you realize that solving the mystery of the unknown mixtures was like doing a logic problem? Ask your teacher to give you a new logic problem to solve with just a pencil and paper.

Figure 14-2

*Testing the
unknown
mixture*

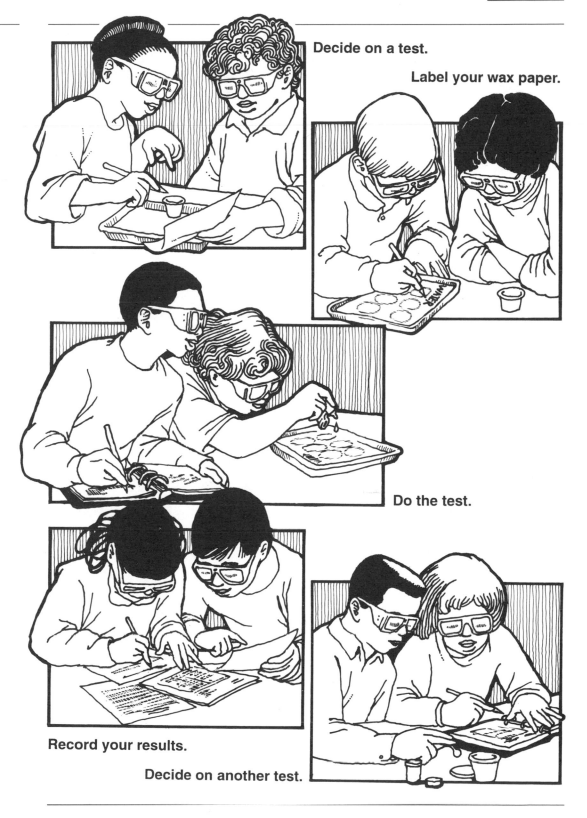

Decide on a test.

Label your wax paper.

Do the test.

Record your results.

Decide on another test.

Testing Household Liquids with Red Cabbage Juice

Think and Wonder

Remember the red cabbage juice test from Lesson 9? What do you think will happen if you mix the juice with other liquids? In this lesson, you will add the juice to six liquids you might find in your home. What changes will occur? What else can you discover about the red cabbage juice test and your five unknowns?

Materials

For you

   1 science notebook
   1 **Record Sheet 15-A: Household Liquids Test Results Table**
   3 strips of colored construction paper

For you and your partner

   1 science pail
   1 sheet of wax paper
   1 dry-erase marker
   1 tray
   2 paper towels
   1 dropper bottle of red cabbage juice

For you and your team

   6 labeled dropper bottles of household liquids (water, lemon juice, detergent solution, vinegar, ammonia solution, alcohol solution)
   1 tray

Find Out for Yourself

1. What did you find out about red cabbage juice when you added it to the five unknowns in Lesson 9? Share your thoughts with your class.

2. Look at your teacher's list of six household liquids. What might happen if you added red cabbage juice to each of them? Record your ideas in your notebook. Then share them with the class.

3. When your teacher gives you **Record Sheet 15-A: Household Liquids Test Results Table,** go over the instructions on pg. 68 for testing the six liquids. Pay close attention so you'll know what to do.

4. Pick up your materials. Your teacher will give your team the tray of six liquids to test.

5. Following the instructions, start testing the liquids.

6. Clean up. Your teacher will collect the tray of household liquids.

7. Referring to Record Sheet 15-A, share your results with the class.

8. Your teacher will give you three strips of colored paper. Review your red cabbage juice test results for each liquid. As your teacher names a liquid, silently hold up the colored paper closest in color to that liquid's result.

9. Turn to a clean sheet of paper in your notebook. Draw three large circles. Try to divide the liquids into different groups. Base your groups on the color each liquid turned when mixed with red cabbage juice.

10. Share your groups with the class. Give reasons for your decisions.

11. Read "The Case of the Disappearing Stomachache," on pg. 70.

12. Discuss the story with your class.

    ■ What did you learn about chemicals by reading the story?

    ■ Which of the six household liquids have properties similar to those of the chemicals in the reader?

    ■ How are the properties similar?

    ■ What do you think the red cabbage juice test can tell you about the six liquids?

13. Look back at the groups of liquids you drew in your notebooks. How can you now group the six household liquids into acids, bases, and neutrals?

14. Help your teacher decide whether each of the six liquids is an acid, a base, or a neutral. Then discuss the properties of acids, bases, and neutrals with the class.

15. How can you use this new information about chemicals to discover more about the first five unknowns you tested? Talk about this with your class.

16. Take out your "Test Summary Table." With a partner, look at your results from the red cabbage juice test. In your notebook, answer these questions:

    ■ How did each of the original five unknowns react with red cabbage juice?

    ■ Using what you just learned about acids, bases, and neutrals, what can you now say about the original five unknowns?

    ■ Where would you put each unknown on the class board of acids, bases, and neutrals?

17. Share your discoveries with your class.

18. Put your record sheet and "Test Summary Table" back in your notebook.

Idea to Explore

With your teacher, set up a classroom learning center where you can test other household products with red cabbage juice. Some interesting products to test are shampoo, toothpaste, soap, juices, ginger ale, and milk.

Student Instructions for Testing Household Liquids with Red Cabbage Juice

1. Set up your tray with the test mat and the wax paper over it.

2. Choose a household liquid to begin your test. Record its name on **Record Sheet 15-A.**

3. Draw a seventh circle on your test mat to use as your compare circle. Put three drops of the household liquid in the compare circle.

4. Using the dry-erase marker, write the household liquid's name above the first test circle. Put six drops of the liquid in the circle.

5. Add six drops of the red cabbage juice to the liquid you are testing.

6. Observe the mixture for about 15 seconds and record what happens on **Record Sheet 15-A.**

7. With a paper towel, remove the liquid from the compare circle.

8. Repeat Steps 2 through 7 for the five other household liquids.

Reading Selection

The Case of the Disappearing Stomachache

Has this ever happened to you? You take a big bite out of a sour pickle. You like it so much that you eat three more. Chances are, you get an awful stomachache. So, you take some stomach medicine. Soon, your stomach stops burning and you feel better. What's going on? Here's a clue. It has to do with two groups of chemicals: acids and bases.

Acids are found in foods like lemons, clear sodas, apples, vinegar, and, of course, pickles. Most of these acids taste sour and have a strong smell.

Then, there are the kinds of acids you **don't** eat, because they are poisonous. Some of these are used in fertilizers, polishes, and car batteries. Many of these acids are so strong that they can burn your skin or clothes.

Bases are found in detergent, oven cleaner, cement, baking soda, bleach, and the pills you take to make your stomach feel better. Some bases have a bitter taste and some burn. They're often slippery like soap.

Acid or Base: How Can You Tell?

How can you tell whether a chemical is an acid or a base? In the 1600s, a scientist named Robert Boyle did some experiments using the juices from plants such as violets and roses. When he added acids to the plant juices, they turned colors— either pink, red, or bright purple. When he added bases to the juices, they turned green.

We call these juices (like the red cabbage juice you used in your tests) **acid-base indicators.** Why? Because the juices indicate (or tell), by a change in color, whether a chemical is an acid or a base.

What about the chemicals that don't turn these plant juices pink or green? We call these substances **neutrals.** Neutrals—like water—are not acids or bases. But when you mix the right amounts of an acid and a base, you get a neutral substance. That process is called **neutralization.**

Why Did Your Stomachache Disappear?

So why did your stomachache go away? The pickles you ate caused too much acid to build up in your stomach. And the stomach medicine is a base.

When you swallowed the medicine, it mixed with the acid in your stomach and neutralized it. And you felt much better. (Next time a bee stings you, have an adult put some baking soda on it. What do you think will happen?)

Now, go back to your chart of results from the red cabbage juice tests.
What are the acids?
What are the bases?
What are the
neutrals?

Using the Known Solids to Identify Unknown Liquids

Think and Wonder

A dropper bottle contains one of the liquids you used to test the five chemical unknowns. But the color of the liquid is disguised. Which liquid is it? How can you find out? In this lesson you will solve the mystery of the unknown liquid.

Materials

For you

    1 science notebook

    1 **Record Sheet 16-A: Unknown Liquids Test Results Table**

For you and your partner

    1 science pail

    1 tray

    1 test mat

    5 toothpicks

    1 bottle of unknown solution A, B, or C

    1 dry-erase marker

    1 sheet of wax paper

Find Out for Yourself

1. Your teacher is holding up three bottles of liquid. You've used all of these liquids in your chemical tests. How can you find out which liquid is in each bottle? Share your ideas with your class.

2. Go over **Record Sheet 16-A: Unknown Liquids Test Results Table** with your teacher.

3. Ask your teacher any questions you have about testing the unknown liquids and which materials to use.

4. Pick up your materials.

5. When your teacher gives you one of the unknown liquids, you and your partner can start testing. Remember to record the unknown liquid you were given (A, B, or C) on your record sheet.

Figure 16-1

Testing unknown liquids

6. Clean up.

7. In your notebook, record your answers to these questions:

   ■ Which tests gave you the most information?

   ■ Which properties helped you identify the unknown liquid? Why do you think these properties were so helpful?

   ■ How did you use negative results to support your conclusions?

8. Share your test results with your class. Tell them which liquid you and your partner think you have. Explain why.

9. Hand in Record Sheet 16-A to your teacher.

Ideas to Explore

1. Do you want to solve another chemical mystery? Ask your teacher if you can test one of the other unknown liquids.

2. With your class, create a "Chemical Information Sheet" (like the one you used in Lesson 12) on the unknown liquids. Using the properties of each liquid, write a "What Am I?" description. You can include the red cabbage juice, too.